TH1
EVOLUTION
CRISIS

5

Five Evolutionests

Think Again

PENFOLD BOOKS

Contents

Introduction

Charles Darwin's 1859 book, *The Origin of Species*, radically reshaped the world's view not merely of science, but of politics, morality, sociology, psychology and religion.

Yet as the 21st century dawned more and more scientists were expressing their doubts as to Darwin's central theory. Over 700 have declared, *"We are sceptical of claims for the ability of random mutation and natural selection to account for the complexity of life. Careful examination of the evidence for Darwinian theory should be encouraged."* (See www.dissentfromdarwin.org).

In 1976 Professor Dean Kenyon repudiated the conclusion of his own evolutionary University textbook *Biochemical Predestination* (1969) which he had co-authored with Gary Steinman. His intensive research of amino acids and DNA caused him to reject Darwin's theory and accept Intelligent Design. Then, in 1985, the atheist, Dr. Allan Rex Sandage, regarded as the greatest observational cosmologist in the world, told an American conference on science and religion that he had become a Christian, declaring:

"It was my science that drove me to the conclusion that the world is much more complicated than can be explained by science...It was only through the supernatural that I can understand the mystery of existence...Many scientists are now driven to faith by their very work."

It is our hope that this book about five scientists' journeys from Darwinism to creationism will be used to raise questions about evolution and challenge readers to look at other options.

CHAPTER 1

The Challenge of Creation

Dr. A.J. Monty White
Market Harborough, England.

Dr. A.J. Monty White holds a B.Sc. (Hon.) in chemistry and a Ph.D in chemistry (in the field of gas kinetics) from the University of Wales, Aberystwyth. During his academic career he held a number of senior administrative positions at Cardiff University.

M y parents could hardly have been described as religious. Father was an atheist; mother an agnostic. During childhood I was told that God did not exist, Christ was not the Son of God, and the Bible was nothing more than a collection of fairy stories.

As with most British school children, church attendance made little impression on me. To the contrary, it put me off Christianity altogether. The church's vicar seemed to have little time for families from poor backgrounds. He never encouraged me to read the Bible and nothing he ever said or did in any way convinced me that God existed. About all I remember were his purple robes and the boring read prayers!

At age eleven I passed an examination that enabled me to attend the local Grammar School. I stopped attending church and gradually developed into a protagonist for atheism, just like my father before me. I rejected the Bible's claim to divine authorship, dismissing its predictions of future events as so vague that one could read almost anything into them. The New Testament's record of the life of Christ sounded fanciful to me, especially his supposed resurrection from the dead, not to mention the incredible stories of miracles dotted throughout the rest of the Bible. I simply could not have any intellectual respect for such a book. After all, if Christianity was such a miraculous religion, why weren't miracles happening today?

In October 1963 I left home to pursue a degree in chemistry at the University College of Wales, Aberystwyth. To make friends, I attended various meetings of the Christian Union and one of the church societies. The Christians I met at these events differed greatly from those I had known previously. They didn't just believe in God – they claimed to know Him! I raised my usual objections to the Christian faith but it wasn't long before I encountered some solid answers to my arguments. In particular, I discovered that the predictions in the Bible were actually much more accurate than I had thought. For instance, in Micah 1:6 the Bible predicted the destruction of the city of Samaria:

"Therefore I will make Samaria a heap of rubble, a place for planting vineyards. I will pour her stones into the valley and lay bare her foundations."

There are four specific predictions in this verse. Firstly, Samaria's ruins will become a heap of rubble. Secondly, the stones used to construct Samaria will be pushed into a valley. Thirdly, Samaria's foundations will be laid bare and fourthly, Samaria will become a place where vineyards are planted. This prediction was made around 730 B.C., but it was not until 1265AD, almost 2,000 years later, that the prophecy was fulfilled – thus no one can argue that it was written down after the event. As the Bible predicted, Samaria was totally destroyed and has never been rebuilt. The ruins were cleared away in order to use the site for agricultural purposes, and in doing so Samaria's foundations were dumped in a nearby valley and grapevines ended up growing on the site. But my scepticism wasn't about to be blown away by one verse. After all, a naturalistic explanation was bound to turn up if I looked hard enough.

My friends also claimed that Jesus had fulfilled many ancient prophecies during his life, yet this claim was based on 'evidence' from the Bible – how convenient! As far as I was concerned, the gospel writers had merely fabricated their account to make it appear that these prophecies had been fulfilled. My friends responded by claiming that the New Testament was a literal and accurate account of what really happened.

To get to the bottom of the question of the reliability of the New Testament I turned to a book called *The New Testament Documents: Are They Reliable?* by the late Professor F.F. Bruce, Rylands Professor of Biblical Criticism and Exegesis at the University of Manchester. In this detailed work, Professor Bruce provided ample objective evidence for the authenticity, historicity, truthfulness and reliability of the New Testament writings. Faced with these facts and others raised in discussions with my friends, I began to see that I had no real alternative but to accept that the New Testament documents were historically accurate and had neither been doctored nor altered over the centuries to fit a biased Christian perspective.

Within a few months of going to university, I found that all my arguments for not believing in God had evaporated: I was convinced that the writings in the Bible, especially those of the prophets, could be trusted; that Jesus Christ was who he claimed to be and that he did rise from the dead. I even accepted anecdotal evidence from my friends that miracles did happen today, though I never experienced a miracle myself. Still, my intellectual assent to all of this never impacted my life until February 1964 when I experienced a real conversion to Christ, as I repented of my sins and discovered the joy of being forgiven through trusting the Lord Jesus as my redeemer and saviour.

The following October, I began to study geology at university. My first lecture was given by Professor Alan Wood, my Head of Department. He launched into an explanation of how inorganic chemicals on the earth's pre-biotic surface combined to produce organic molecules that formed themselves into self-replicating organisms, which in turn evolved into all the life-forms that have ever existed on the earth. Professor Wood was also at pains to point out that *Homo sapiens* was not the end product of evolution. He suggested that in a few hundred million years, others would find fossilised remains of 20th century humans and declare, *"How primitive!"*

I left the lecture in deep thought. How could I reconcile Professor Wood's account with what the Bible teaches in Genesis about the creation and the early history of the earth? I decided to ask my Christian friends about the creation/ evolution question. I was surprised at their responses. They basically all told me to simply accept evolution and to interpret the early chapters of Genesis accordingly. Such an idea is called theistic evolution: evolution has occurred but has been controlled by God. The early chapters of Genesis are not to be interpreted as history, but as myths, allegories and legends. Feeling I had really no other choice, I bought into this concept and held to it for a number of years.

I continued to thoroughly enjoy my geology studies. Within two years I had advanced to degree level. I also continued to major in chemistry, obtaining an honours degree in that subject in 1967, thereafter beginning research for my doctorate in the field of gas kinetics. During this time I married and shortly afterwards my wife began challenging my views on theistic evolution. She asked me to explain a verse found in 1 Corinthians 15:22, *"As in Adam all die, even so in Christ shall all be made alive."*

I realised that I was being asked to answer the fundamental question, 'Who was Adam?' I remember thinking that if I believed in a literal Adam, I would also have to believe in a literal Eve, a literal garden of Eden and a literal six-day creation. In short, I would have to commit intellectual suicide, for at that time I knew not a single academic who believed any of these things. Everyone I knew accepted evolution as a fact. Every book I read, even those written by Christians, taught evolution. What was I to do?

To answer my wife's question, I re-read the New Testament to see what its writers and characters thought of the early chapters of Genesis. I soon discovered that in the New Testament, the creation, Adam and Eve, the fall, Noah, the flood and so on – are all accepted as literal and historical. Therefore to hold to the Bible would mean abandoning evolution. However, over the next two years, I came to the conclusion that it was possible to reject the idea of evolution and accept the historicity of the early chapters of Genesis without committing intellectual suicide. I wrestled with this issue the whole time I was busy pursuing my academic studies.

My investigations centred on three major areas: chemical evolution, the fossil record and dating methods. As study tools I used my old geology lecture notes and various scientific textbooks. At the time, I did not know a single creationist and I had never read a single anti-evolution book or article. Strange as it may seem, I became a creationist as a result of reading pro-evolution material!

I was amazed at the naivety of the statements made about so-called chemical evolution. Evolutionists purport to have proven, by various experiments, that life originated by chance on a pre-biotic earth. Yet their experiments are designed not by chance, but by intelligence! What in fact they are saying is, "If we can synthesise life here in the laboratory, we will have proven that no intelligence was needed to create life in the first place!"

In Stanley Miller's famous 1953 so-called 'origin of life' experiment, a number of amino acids were produced by passing an electric discharge through a mixture of ammonia, hydrogen, methane and water vapour. Since that time, various mixtures of amino acids, sugars and nucleic acid bases have been produced in similar experiments. Since these chemicals are the building blocks of living systems, it is argued that such experiments prove beyond doubt that a creator was not necessary for the origin of life – it could have happened by chance. However, in Miller's experiment, amino acids were successfully produced only because they were removed from the mixture as soon as they were formed. Had they been left in the apparatus, they would have been destroyed by the very same electrical discharge that synthesised them. Furthermore, as in all such experiments, the amino acids were produced in both right-handed as well as left-handed forms, whereas living systems contain only left-handed amino acids. Additionally, had oxygen been present in the mixture of gases, the amino acids would never have formed at all. This point is extremely important because evidence from geology indicates that the earth's atmosphere has always contained oxygen. Hence the mixture of gases in Miller's experiment did not represent what evolutionists now believe existed in the early earth's atmosphere. The chemical evolution of life is still a theory born out of dogma, wholly unsupported by empirical evidence. Miller's experiment used the wrong ingredients, employed the wrong methods and produced the wrong results – apart from that it was a brilliant experiment!

The second area I looked into was the fossil record, which I soon realised did not show the gradual evolution of one life-form into another as predicted and demanded by

9

evolution. 'Missing links' are so named because they truly are missing – none has ever been found. Gaps exist in the fossil record at all the major breaks: fish to amphibians, amphibians to reptiles, reptiles to birds and reptiles to mammals. Furthermore, the so-called fossil remains of creatures linking humans to their ape-like ancestors are all open to question. The field is riddled with hoaxes, forgeries and misrepresentations. The famous Lucy skeleton (from the australopithecine group), which is supposed to represent the first creature in the human line as it broke away from its ape-like ancestors, is shown in the Natural History Museum (London) with totally human hands and feet – even though there were no hands and feet found at the site of its excavation. Indeed, all the australopithecine hands and feet that have ever been found exhibit the curved ape-like bones that prove them to be very far from human.

The final scientific area I looked at was the crucial issue of dating. How do we know how old a rock is? As a chemist I could see that the accuracy of all the dating methods depended on a number of assumptions, some of which were unprovable and others unknowable. For example, in order to determine the age of a rock by radiometric dating, three things must be known:

1. The present concentrations of the parent and daughter elements in the rock.
2. The original concentrations of parent and daughter elements in the rock.
3. The rate of decay of the parent into the daughter element.

Now, in most cases, it is possible to measure the amounts of parent and daughter elements currently existing in the rock (1). However, it is not always possible to know the original concentrations. Sometimes it is assumed that there was no daughter element present when the rock was formed, though there is no way of proving this. It is an assumption.

10

Again, though the present rate of decay of parent into daughter can usually be measured accurately, there is no way of knowing that this rate has not changed over time. It is another assumption.

One proof of the accuracy of the different dating methods would be that they all give the same age for the same rock sample. However, as I researched the literature, I became aware of reports that different methods gave different ages for the same rock. In these papers the authors spent a great deal of time discussing why there were discrepancies and why the age should really be determined from the fossil content of the rock or from the fossils in the adjoining rocks. That of course involves circular reasoning: the age of the rock is determined by the fossils, while the ages of the fossils are determined by the rock!

I finally concluded that evolution was at best an unproven hypothesis. I became convinced that most people believe in evolution because they choose to, not because of overwhelming evidence. In fact many evolutionists readily admit that they believe in evolution not because science has proved it, but because the other alternatives, especially 'God', are totally unacceptable. Personally, after a long struggle, I accepted creation as the best explanation for the beginning of the universe, the origin of life, the information content in our DNA, the laws of the universe and much more. I saw that the fossil record confirms the Biblical fact that both plants and animals reproduce after their own kind – and so, for the last 35 years I have continued to believe and promote creation science and am happy to record my testimony here, in the hope that it may be a help to you the reader.

CHAPTER 2

Testing Truth with an Open Mind

Dr. Roy Spencer

Dr. Roy Spencer is a senior scientist at the Marshall Space Flight Centre, USA, and is a leading scientist with NASA. At one point in his career he was asked to advise the White House on global temperature trends.

To be honest, what little I knew about Christianity bothered me! In particular, Christians in my area who went from house to house inviting people to events at their churches *irritated* me! These people clearly believed they were part of the one, true religion, if indeed there was one. I asked myself, how could they be so sure? If Christianity were true, why weren't most people Christians? How could anyone in good conscience devote his or her life to any one religion without at least investigating all the other world religions too?

I also had a fundamental problem with the Bible. Was not its first book, Genesis, merely a mythical account of how the universe and life came into being? Anyway, it seemed to me that Christians picked and chose what they wanted to believe, selecting some things in the Bible, while rejecting others, often quite arbitrarily and subjectively – how, then, could I regard their 'holy book' as the inspired Word of God? Fundamentalists were all a bunch of biased faith-heads, while scientists were objective, honest, unbiased and open-minded. Well, sort of! The time came when I began to realise, to my initial surprise, that there was a group of scientists who believed that the universe and all life within it had been created by some greater intelligent Being, not by mere chance. They were seemingly able to do so using scientific arguments, not just religious dogma. I began to study their case and after some months of analysis I finally became convinced that the theory of creation actually had a much better scientific basis than the theory of evolution, for the creation model was actually better able to explain the physical and biological complexity in the world.

The possibility then presented itself that, despite all I had previously thought, Genesis, the first book of the Bible, might actually be true! That realization led me to open a Bible for the first time, and to read it for myself, from the beginning. I also became open to reading the Bible because I discovered that a very intelligent friend of mine believed the Bible was the word of God. My family and I accepted this friend's invitation to accompany him to church one Sunday. I was impressed by the genuine concern and friendliness shown by many of the people. Clearly Christianity was not a faith confined to simple, or even socially maladjusted, people, as I had previously thought.

As I investigated religions other than Christianity, I became aware that many of them assume evolution to be true. The Bible was the only 'holy book' in which I could find a record of God's creating the material universe from nothing! Next, the work of many historians revealed to me that the Bible is by far the most accurate and best-substantiated ancient book known to man. It truthfully portrays actual historical events and has been faithfully copied by scribes over the centuries so that what we have today in the Bible is, to a very high degree (within a percentage point or two), known beyond a shadow of a doubt to be the same as was originally written down by the authors. Furthermore, nothing in that two percent affects any of the major Bible teachings or events.

When I turned to the gospels I learned that the contemporary enemies of Jesus, who wanted to disprove His divinity, could not deny His many miracles, there being too many eye witnesses. Not being able to dispute the *fact* of His amazing deeds, they questioned the *source*: they asserted, feebly, that an evil superhuman power had performed the miracles, not the Spirit of God!

I was struck by the unity of the Bible's message – the way it agreed with itself even though it was written by 40 different authors over a period of 1,600 years. I realised that the gospel records were free of comment from the writers. They merely recorded what they saw without exaggerating the events, without covering up the faults and failings of the followers of Jesus and without trying to present the story in exactly the same way. There were enough differences between the four gospels to prove they had not collaborated, but not enough differences to stray into the area of outright contradictions and errors.

So, at last, I had to face the reality, based on all the evidence, that the basic tenets of Christianity were true, and that the gospel of Christ really changes people's lives. True, my decision to become a Christian involved faith, but not the kind of faith caricatured by the likes of atheist Richard Dawkins, a faith that 'just believes' in the teeth of real evidence to the contrary. My faith in Christ was evidence-based. I had very well founded reasons for believing in Him. In fact, the eyewitness writer of the fourth gospel, John, explains why he recorded what he saw – *"That you might believe that Jesus is the Christ the Son of God "* (John 20:31). So, having examined the Biblical record of creation and the person and work of Christ on the cross for my sins, of which I knew I was guilty, I put my trust in Him for salvation and, to explain to others what had happened to me, I was baptized.

To examine the relationship between science and the Bible, a good place to start is with the origin of the universe.

14

Science presents us with the laws of thermodynamics, the first of which states that the total amount of matter and energy in existence is constant. If this were the only natural law to be satisfied, it would be possible to believe that the universe has existed forever. Indeed, that was the prevailing view back in Darwin's day. However, the second law of thermodynamics states the overall amount of *useable* energy is constantly decreasing – it is being degraded into a less useful form. In other words, the universe is dying. If the universe were eternal it would by now have experienced what astronomers call a 'heat death' – a state of total equilibrium in which entropy would be infinite. This, among other factors, has led a majority of astronomers to agree that the universe had a beginning. Several thousand years of scientific endeavour has brought the majority of scientists in line with the first verse of the Bible which states, *"In the beginning God created the heavens and the earth."* Well, the first three words of the verse anyway! If there is no God, who or what caused the universe to begin? There really are only two basic options – it created itself out of nothing or it was created by something greater than itself! If everything that *begins* to exist has a cause, and the universe began to exist, the universe must have had a cause.

A second issue is the origin of life. There is a vast gulf between the most complex non-living compound such as a crystal and the simplest form of life such as a bacterium. The gap is much larger than the gap between a bacterium and a human being. Science, despite expending enormous amounts of time, is actually further away from an explanation as to how non-living chemicals can accidentally and spontaneously come alive than it has ever been. All the evidence on hand, both in nature and in the laboratory, points to the fact that life only ever comes from life. The Bible credits the origin of life to the power and design of a 'living' creator God.

A third huge issue is the complexity of life. In recent years scientific advances have uncovered the complexity of the cell, both biologically (DNA, RNA, proteins, amino acids, etc.) and atomically (electrons, protons and neutrons etc.). It turns out that the nucleus of every human cell is a digitally coded database containing more information than Wikipedia, and is vastly more complicated than New York City. An increasing number of scientists consider it to be impossible that such a structure could have evolved through random processes, as evolutionists assume. The last 50 years or so have seen real evidence come to light that random mutation and natural selection are incapable of building complexity. Observation of malaria, *E. coli* and HIV, all of which exist in vast numbers and have short life cycles, have shown that while 'Darwinian' processes can cause minor changes, always involving a loss of complexity, they cannot build complexity – nor can they begin to explain where the proteins and genes came from in the first place. Again, the Bible has said all along that life was originally created and has ever since reproduced 'after its kind'.

Science has startled us with its many discoveries and advances, but it has hit a brick wall in its attempt to rid itself of the need for a creator and designer. In fact, every year that passes reveals all the more starkly that a naturalistic explanation for the origin of the universe, life, complexity, consciousness and reason is not merely 'difficult' but hopelessly impossible. It took me a long time to finally approach the Bible with an open mind, but I am very glad there came a time when I did. My advice to you would be to seek out the truth for yourself. Unfortunately, much of what people believe is based less on evidence and more on unsubstantiated just-so stories. In relation to the basic claims of Christianity, do what I did! Read the Bible. Judge it for itself. Put it to the test. I am confident that you too will find the Bible not only to be in agreement with proven facts of science, but also to be the book which will lead you to a personal faith in God the creator of all things.

CHAPTER 3

From Oxford Atheist to Leading Creationist

Professor Arthur E. Wilder-Smith

A.E. Wilder-Smith F.R.S.C., Ph.D was a Countess Lisburne Memorial Fellow in cancer research for London University and a member of the British Chemical Society, the American Association for the Advancement of Science and the New York Academy of Science. He held the Chair of Professor of Pharmacology at the University of Illinois, Chicago, where he was elected 'Best Teacher' four times and won the 'Golden Apple' award three times.

Born in England in 1915, Arthur Wilder-Smith grew up as the eldest son in a well-to-do farming family. An inquisitive child, he went to Oxford University in 1933 to study botany, zoology and chemistry under Professors de Beer, Ford, Robinson, Chattaway and others. There his atheism became firmly entrenched, much to the grief of his devoutly Christian mother.

In 1936, at the age of 21, Wilder-Smith's life took an unexpected turn. A series of events conspired to radically challenge his atheistic worldview. In his own words:

"At this time, a young English General moved into our part of Berkshire. He was the youngest General in the British-Indian army and had been stationed for some years on the North West frontier between India and Afghanistan. He was a genuine soldier: straight, fearless, intelligent and utterly honest. After his conversion to Christ at the age of 45, he decided without delay that he would use the rest of his life in service for his new 'King'. He took early retirement and bought a house on the Thames. There he built a small chapel on the lawn, there being no active congregation nearby, and gave evangelistic sermons every Sunday. Everywhere he went, he held services – in churches, chapels and community halls.

"My dear mother was invited to hear him. She went and, being suitably impressed, invited her sister, Aunt Addie, to accompany her, who promptly became a believer along with a cousin of mine. Then my mother extended an invitation to me. I told her politely but firmly that an Oxford University student does not go to evangelical meetings, not even if a General is speaking! My mother poured out her frustration with me to the General personally, who said cryptically, 'When Mohammed will not come to the mountain, then the mountain must go to Mohammed!' He invited me to a high tea at his home. Now in polite English society, when a General invites a person to high tea, he has no option but to go. To refuse is socially unacceptable.

"So, on a lovely afternoon, I drove to 'Watersmeet' and received a friendly welcome. We played tennis, rowed a boat on the Thames and had tea with the family. The General and I retired to his study to talk privately. He asked me if I was a Christian. I answered that I was a committed atheist, even though I had been baptized and confirmed. He said that he was a committed Christian and believed that Christ had died for his sins and risen again. I laughed at his naiveté and asked him how such an educated man as he could believe in the fairy tales of the Bible. Christ is recorded in the gospels as believing in Adam and Eve, Jonah and the whale, and the worldwide flood of Noah. No educated person today could believe in such nonsense. Christ clearly did not know the difference between history and myth. On that basis alone he could not possibly be the 'Son of God'.

18

"During the course of the conversation, I told him that Darwin's theories were nothing more or less than the hard facts of history and natural science. The world as we know it had come about through chance and natural selection. The idea of an intelligent creator was the ultimate in unscientific thought; and anyway, the theologians of today no longer believed in creation. There was no proof of the existence of God: Feuerbach and others had long since proven that. The General's religion was uneducated fantasy and imagination.

"The honourable General, who actually was not at all uneducated, looked at me with sadness. He admitted that he knew next to nothing about natural sciences and that he had not made much progress with me. Around 11.00pm I said good night. I was sure I had well and truly beaten him. His dear wife, Mrs. Frost (a true lady), later told me that she had found me unbearably conceited and had advised her husband to give up on me. I was absolutely and hopelessly 'unconvertible'. However, the General had a weapon of which I knew nothing. He understood the power of intensive prayer. So, for three weeks, he prayed for me after which he again invited me to tea. This time I had no reservations about the visit. The house on the Thames was beautiful, I enjoyed rowing on the river and the tennis court was good fun. If need be, I could easily dispense with the General and his arguments. My education had instilled in me an attitude of arrogance. I am ashamed to say that I did not at the time appreciate how unwise, rude and even naïve my actions and thoughts were.

"Late in the evening we again retired to his study to converse. His experience as a General had made him a man with obvious strength of character and authority – a man with a gift for command. This time he used a fresh strategy with me. He began, not with natural science, but with personal character and self-discipline. Little did he know how close to the bone he was. I was a frustrated young man. I lacked motivation, was often despairing, not to mention bad-tempered. It was plain to me that the General was a gracious man whose whole life was reflected in his manner. His serenity and strength shone in his eyes. His whole demeanour and bearing had an effect on me as he spoke. His words made a deep impression on me, each one being underlined by his character. I felt hollow in the light of his solidity, shallow in the face of his depth.

"Make no mistake. The evening was no emotional trick. The General did not mince his words. He made it clear to me with great evangelical solemnity that my sins, my violation of the eternal law of God, had ruined me. I could have taken offence but for the fact that I knew what he was saying was true and I was keenly aware of his genuine, loving interest in my eternal welfare.

"After a long conversation, he asked me directly if I felt my need of forgiveness and the transforming power of God in my life. My answer was: "Yes. *A thousand times, yes.*" He then suggested we both kneel down. I heard him pray out loud for me. When it was my turn, not a word would come to my lips. I was speechless! While still on our knees, the General asked whether I believed that the Lord Jesus Christ lived in his own heart. That I could not deny. Did I desire above everything else to have the same Spirit in my heart? To that again, I resolutely answered yes. Do not think the General tried to manipulate me like so many evangelists do today who only wish to achieve results and statistics for their mission societies. He spoke man to man. I felt a sense of ease when I was with him. His method was one of cogent, wise and Biblical encouragement – not emotional pressure. That day God enabled me to call out to Him for the forgiveness of my many sins. I understood that Christ had died for me and on the basis of His finished work on the cross accepted forgiveness by faith. When I stood up again the sceptical General asked me if I was ready to confess Christ to others. "*How would it be, if you go into the kitchen now and tell my wife and children what has just transpired here?*""

"I hesitated because his children were either older than me or were my own age and because I had noticed that Mrs. Frost, though a perfect lady, held me at somewhat of a distance. Yet I did not want to be a coward in front of such a General – especially not in front of his family! So, I briefly related my conversion to them and by thus overcoming my own inhibitions, pride and fear, joy flooded my soul. Mrs. Frost was simply overjoyed. *"The days of miracles are not yet passed after all!"* she exclaimed, and cordially shook my hand.

"Though now a Christian, my intellectual difficulties about Adam and Eve, evolution/creation and the miracles in the Bible were not totally cast off. I regularly discussed these problems with the good General but he could not really help me. He felt I must 'just believe' and everything would be all right. However, one comment he made did help me. He remarked that often when we trust the Lord and His Word even when we do not understand, God sends people across our path who are in a position to answer our questions more satisfactorily."

After his conversion, Wilder-Smith earned a PhD in organic chemistry at Reading University, followed by another doctorate in biochemistry from the University of Geneva and one in biology and natural sciences from E.T.H. in Zurich. Ultimately it was through his own scientific investigation, rather than discussions with Christians, that he came to reject Darwin's theory of evolution and began to promote creationism. He eventually became the Professor of Pharmacology at the University of Illinois, Chicago, and his work led him all over the world, speaking about creation and other issues in hundreds of lectures in many famous universities. Fluent in German, he was particularly effective in reaching many thousands of German POWs during and after the Second World War. He was also a big hit with students who appreciated his boundless patience when they barraged him with questions.

"Today in Europe and the USA, the teaching of evolution in the schools and universities is a great problem," he said. "It is taken today as an incontrovertible fact of science that Darwin has made the idea of a divine Creator superfluous for the educated person. If God is scientifically superfluous to creation, then Christ who called Himself One with the Creator God, automatically becomes superfluous too. Thus, since Darwin, the preaching of Christ, particularly in academic circles, has become increasingly lacking in urgency."

Wilder-Smith authored many books on biogenesis, his particular field of expertise. (A list of his books can be seen at www.wildersmith.org). In *The Natural Sciences Know Nothing of Evolution* he exposed the idea that life arose spontaneously from the ocean:

"...if excess water is present in the reacting mixture, peptide synthesis does not take place, equilibrium remains on the side of the initial reagents, the amino acids, which are the building blocks of life. This phenomenon is covered by the law of mass action: it is valid for all reversible reactions. Briefly said: in reactions of this type, synthesis of polypeptides from amino acids does not take place in the presence of excess water. The consequence of this well-known fact of organic chemistry is important: concentrations of amino acids will combine only in minute amounts, if they combine at all in a primeval ocean providing excess water, to form polypeptides. Any amounts of polypeptide which might be formed will be broken down into their initial components (amino acids) by the excess of water. *The ocean is thus practically the last place on this or any other planet where the proteins of life could be formed spontaneously from amino acids.* Yet nearly all textbooks of biology teach this nonsense to support evolutionary theory and spontaneous biogenesis...Has materialistic Neo-Darwinian philosophy overwhelmed us to such an extent that we forget or overlook the well-known facts of science and of chemistry in order to support this philosophy?

"Approximately twenty amino acids comprise the basic building blocks of life from a material point of view. Without these, life as we know it today could neither originate nor exist. Some of these amino acids can, under certain circumstances, be formed in the primeval atmosphere through chance lightning, as we have already discovered. But to state, as many experts do, that these amino acids which are formed by chance can be used to build living protoplasm is certainly grossly erroneous in principle, for they are for such purposes, in fact, entirely useless. Without exception all Miller's amino acids are completely unsuitable for any type of spontaneous biogenesis. *And the same applies to all and any randomly formed substances and amino acids which form racemates.* This statement is categorical and absolute and cannot be affected by special conditions."

In his book *God: To Be or Not to Be?* he wrote about DNA: "Once we have reached the arrangements of matter of complexity such as DNA molecules the sailing is fairly plain. But we have no known way of accounting for the original order of life residing on DNA or the enzyme systems producing it, which must have come from some source apparently outside matter and able to convert energy into codes and sequences...Why should it be anathema to Monod and the materialistic scientists to deny *a priori* any source of information/energy conversion outside matter?

"May not thought itself be the source of material order, sequences and codes with which life is inextricably interwoven? Jeans thought so. For thought consists of concepts embedded in sequences and codes which are, in our experience, imposable on to matter in the form of voice, printed text, poems, song or even memory macromolecules. The matter the poems ride on (paper) is not part of the poem or even the thought behind it. It is merely the medium on which the poem code is simulated and nothing else. We come then to the suggestion that thought, which is in itself not material, but which can ride on matter (paper, grey matter, magnetic tapes, etc.) imposed itself onto amino acid units and their sequences as a 'written code' which bears life and its meaning...Why should it be unscientific, then, to suggest that something similar to human thought, which is a converter of energy into sequences and codes, produced primeval life? Since life's order is not, as far as we can see, present on matter endogenously today, it certainly could not have been resident in or on matter at the beginning (for in the beginning matter was by definition identical with present-day matter.) Thus the primeval orderer must have resided outside matter. Which is merely a polite way of saying that we are forced to conclude that the primeval source of order must have been metaphysical and have resided extra-materially."

In 1985 Professor Wilder-Smith was invited to present the scientific case for creation at England's foremost debating society, the Oxford Union, under the auspices of the British Association for the Advancement of Science. The debate took place on 14th February 1986, as a kind of re-run of the 1860 Wilberforce-Huxley debate. Professor Wilder-Smith and Professor Andrews (University of London) were selected to debate the evolutionary professors Richard Dawkins and John Maynard-Smith. In his part in the debate Professor Wilder-Smith deliberately gave only scientific evidence as to why Darwinism cannot answer the origin of life puzzle, and why the evidence points firmly in the direction of an outside designer using information to order matter. In a direct reply to Wilder-Smith's speech, Professor Maynard-Smith acknowledged that while Darwin had not answered the problem of the origin of life, any creationist pinning their hopes on the riddle of the origin of life as proof that there must be a God would be 'crazy' because, *"Before you're very much older it's going to be solved. I mean, really, you'd be mad to say 'I believe' because scientists can't explain the origin of life. Things are moving very fast in that field."*

As of this present time, well over 20 years since the debate, evolutionary scientists are further away than ever from answering not only the question of the origin of life, but also the origin of consciousness and indeed, the origin of the universe itself. Wilder-Smith's arguments have stood the test of time and the case for an intelligent designer is stronger than ever.

CHAPTER 4

Superb Design

Professor David H. Stone

Dr. Stone is associate professor of electrical and computer engineering at Michigan Technological University. He holds a B.S. and an M.S. in physics from Michigan State University, a Ph.D in mechanical engineering from Michigan State University and an M.B.A from the University of Phoenix. He served 20 years in the United States Air Force on a variety of research assignments and received the U.S. Air Force Research and Development Award in 1986 for contributions to high power microwave system testing.

The tortuous journey toward my present Christian faith began in a traditional church, detoured into atheism, turned abruptly to a simple faith in the Saviour, and finally settled on a solid biblical foundation, recognizing that the Word of God is fully trustworthy, consistent and perfect, both theologically and scientifically.

As I grew up on the south side of Chicago, I was fully engaged in religious activities, but I felt I had no foundation. In short, I was an eager churchgoer but not a Christian. I trusted that some combination of prayers, active service, and avoidance of 'big sins' would earn me a berth in heaven. I didn't realize that all sins are 'big' enough to earn hell, *"For whosoever shall keep the whole law, and yet offend in one point, he is guilty of all"* (James 2:10).

There were two powerful forces working against my belief system. Although all of my immediate family and other relatives were religious churchgoers, my dad was a sceptic.

He took delight in pointing out inconsistencies in church doctrine and in the bloody history of what has often purported to be Christianity – most notably the Inquisition. (What I didn't realize was that true Christians were always on the receiving end of such persecution). The second force was the culture of evolution in which I was immersed. I spent considerable time in the museums in Chicago, which are completely saturated with naturalistic explanations for life. Additionally, everything I read in books or saw on TV that touched the subject of origins was evolutionary. Throughout my youth I was unaware of any evidence supporting creationism – indeed back in the 1960s there was little available on the topic.

As a precocious 13-year-old, I brought a flock of questions to one of the senior leaders in my church on a particular Saturday. He couldn't give me an answer for any of them. That unsatisfying meeting brought my doubts out into the open. The Church didn't make sense, had no answers and since everything in the universe could be explained by evolutionary science, who needed God?

I was a miserable atheist for the next three years. What point is there to life if we are just animals and death means the end of it all? At the depths of my depression, though, God had mercy on me and sent me a friend who was a Christian. He and his family befriended me. I saw the love of Christ in their lives and felt drawn to Him. The truth of John 6:44 resonates with me: "No man can come to me, except the Father which hath sent me draw him: and I will raise him up at the last day." Did God ever draw me to Himself as a 16-year-old! I praise the Lord for His wonderful patience.

I didn't understand how to reconcile evolution with my new faith, so I began to study the subject over the next few years, concluding in the end that true science was perfectly consistent with the Bible. I could accept the truth of Genesis – most notably a six-day creation and a literal worldwide flood

– without compromise. I did not feel there was a need to invent a hybrid theory of origins like 'theistic evolution' to gain acceptance among evolutionary atheists. I could find no satisfaction down the road of imagining that God somehow used evolution over the course of billions of years to produce the world as we know it. What an awful method of creation for a loving God: bloody competition, extinction of millions of species of animals and plants – survival of the fittest and destruction of the unfit! That's not the God of the Bible who provides for the birds of the air (Matt 6:26) and praises those who are kind to animals (Prov 12:10)!

One particular quote from an evolutionist seemed to powerfully confirm my view. Jacques Monod wrote:

"[Natural] *selection is the blindest, and most cruel way of evolving new species, and more and more complex and refined organisms...The struggle for life and elimination of the weakest is a horrible process, against which our whole modern ethics revolt. An ideal society is a non-selective society, one where the weak is protected; which is exactly the reverse of the so-called natural law. I am surprised that a Christian would defend the idea that this is the process which God more or less set up in order to have evolution.*"

As I studied the literature on origins I was excited and impressed to find many serious weaknesses in the arguments for evolution. In fact I would assert that there is overwhelming evidence that Darwinian-type goo to you evolution not only didn't happen but is impossible. Here are a number of reasons why the evidence for superb design speaks against the random processes of mutation and natural selection.

1). The molecules crucial to life are so enormously complex that it is *impossible* for them to arise by chance. For example, a single ordinary protein consists of hundreds of amino acids formed into a precise three-dimensional configuration. These molecules are constructed within living cells with the aid of other macro-molecules such as nucleic acids. There is no way for such molecular complexity to arise spontaneously in nature. In my review of the evolutionary literature I have never been able to find any quantitative discussion explaining how such complexity could have arisen in the wild.

2). The simplest conceivable cell – the smallest possible self-replicating organism – is immeasurably more complex than the most sophisticated designs of human science and engineering. Consider this fact alone: there are about 500 different perfectly regulated chemical reactions associated with the process of metabolism. We may marvel at the complexity of an item like the space shuttle, appreciating the work of the thousands of engineers that goes into its design, manufacture and operation but it is actually easy to demonstrate that a single 'simple' cell is far more complex than the space shuttle or any other human construct.

3). The two cornerstones of Darwinism are random mutation and natural selection, but what can these processes actually do? Recently revealed genetic data on microbial parasites such as malaria, HIV and *E. coli* have provided empirical evidence that even under intense selective pressure, and given an astronomical number of opportunities due to their large populations and short reproductive cycles, random mutation and natural selection can ever only yield trivial, mostly degenerative changes (such as malarial resistance to chloroquine and human sickle cell anaemia which confers resistance to malaria). The fact is, DNA replication in higher organisms includes proofreading to keep error rates at less than one in ten billion. Any errors are almost always neutral or harmful and even when 'beneficial' never lead to the production of *novel* and *meaningful* gene sequences.

4). The genetic variations between different kinds of animals are so substantial that there isn't enough time for them to occur on the supposed evolutionary timetable. Even if mutation and natural selection could produce new and more complex information, it would take many generations for each change to take over a population. For example, because human and chimpanzee DNA are apparently only about 5% different, leading evolutionists cite this fact as possibly the strongest evidence for Darwinism. Even 1% would be quite large across the three billion or so nucleotides that make up the human genome. Human evolution has allegedly occurred over the last five million years. Given a human generation time of perhaps 20 years, only a few thousand significant changes could arise in the population. The numbers don't add up. A few thousand changes cannot account for the millions of differences in the DNA code.

5). The gaps in the fossil record are huge and systematic. Even evolutionists are troubled by these gaps and have come up with another theory called punctuated equilibrium to try and explain it away. Gaps appear between all the classes – invertebrates and vertebrates; between fish and amphibians; between amphibians and reptiles; between reptiles and birds and reptiles and mammals. Add to this the Cambrian explosion – the fact that the deepest-lying sedimentary rocks exhibit billions of fully formed, complex invertebrates across numerous phyla with no trace of their supposed evolutionary ancestors – and you have very powerful evidence against gradual Darwinism. This explosion of fossils remains unexplained by naturalistic mechanisms.

6). Life exhibits an abundance of irreducibly complex systems. Such systems would seem to have no way of working in any partially formed state. The problem for evolution is to explain how such systems could develop incrementally. The scientific literature seems to be empty of any quantitative model that allows for an evolutionary origin for any irreducibly complex system. For example, the cilium used by some cells for locomotion is a complex, self-contained motor using some 200 different proteins for structure and operation. As far as I can tell, no one has ever conceived how a crude evolutionary precursor of this marvel of engineering could possibly function.

7). The sophistication of living design at the organ and at the organism levels is astounding. This is not the mere appearance of design due to natural selection's sleight of hand. A detailed look at specialized organs and optimally designed creatures boggles the mind. For example, the sonar systems of bats and dolphins are wonderfully sophisticated and exceed the performance of any that man has built. According to evolution, *chance* mutations selected by the random conditions of nature have built better and vastly more efficient pumps (the heart), filters (the kidney), rotary motors (the bacterial flagellum), cameras (the eye) and computers (the brain) than will ever be built by *intelligent human agents*.

8). Evolution's advocates support a theory that does not lend itself to the criterion of falsifiability. On occasion, some evolutionists have proposed a test but the results are never favourable. For example, an evolutionist once affirmed that structures like wheels and magnets could not have evolved because they would be useless until fully formed. But these structures have since been discovered in living creatures. The wheel is found in the rotary motor of the bacterium's flagellum and magnets are integral components of the navigation systems of many birds. A creationist is not surprised to find fully developed structures with no hint of incremental development in the fossil record. The most 'ancient' bat fossils exhibit the structures necessary for sonar evidencing that bats have always had this sophisticated ability from the time they were created.

9). I regularly search the evolutionary literature to find the best evidence to support the theory. Much is made of bacteria that evolve resistance to drugs, yet such resistance is always through small deleterious changes, not through the introduction of novel meaningful complexity even of the smallest kind. Take away the drug and the 'normal' bacteria return. Antibiotic resistance that arises from the transfer of bits of DNA among bacteria still does not involve the generation of "new" information, but merely the transfer of existing genetic code. The variations among the beaks of Darwin's finches represent the re-distribution of existing genetic information through sexual reproduction – just as a husband and wife can have children with noses longer or shorter than their parents. The literature is empty of any example of an evolved chain of organisms, transitioning from simple to complex.

10). There is considerable evidence that the universe and the earth are far younger than the billions of years required by evolutionary theory. For example, spiral galaxies would have wound up if they had existed for billions of years. They shouldn't be shaped the way they are! A complex density wave theory of spiral arm formation was invented to try to explain this, but it is not consistent with detailed observations from the Hubble Space Telescope. Additional puzzles about galactic structure have led astronomers to postulate that unobservable "dark matter" will save the day! A simpler explanation would involve a Creator who made the universe fully functional and observable from the start.

I am aware that books have been written on each one of the points above. The intent of this list is to be illustrative rather than exhaustive. As I see my faith, the bottom line is: *"The heavens declare the glory of God; and the firmament showeth His handiwork. Day unto day uttereth speech, and night unto night showeth knowledge."* (Psalm 19:1-2). Wherever I look in nature I see evidence of design and wherever I look in Scripture, I find truth and food for my soul.

CHAPTER 5

From Evolution to Creation

Professor Gary Parker

Professor Gary Parker is a biology teacher who received his PhD in biology with geology at Ball State University, USA, in 1973. He has been admitted to Phi Beta Kappa, an American scholastic honorary, and was elected to the American Society of Zoologists and held a fellowship from the National Science Foundation. He has written five textbooks on biology.

The following testimony which records the fascinating journey of Professor Gary Parker from evolutionary biology teacher to leading creationist promoter and author has been condensed from four radio talks.

Interviewer: Dr. Parker, I understand that when you started teaching college biology you were an enthusiastic evolutionist. Is that so?

Dr. Parker: Yes, indeed. The idea of evolution was very satisfying to me. It gave me a feeling of being one with the huge, evolving universe while continually progressing towards grander things. Evolution was really my religion, a faith commitment and a complete world-and-life view that organized everything else for me, and I got quite emotional when evolution was challenged. As a religion, evolution answered my questions about God, sin, and salvation. God was unnecessary, or at the very least had no more involvement than to originally create the particles and processes from which all else mechanistically followed. 'Sin' was merely the result of animal instincts that had outlived their usefulness, and salvation involved only personal adjustment, enlightened self-interest, and perhaps one day the benefits of genetic engineering. With no God to answer to and no purpose for mankind, I saw humanity's destiny as being in our own hands. Tied in with the idea of inevitable evolutionary progress, this was a truly thrilling idea and the part of evolution I liked best.

Interviewer: Did your faith/trust in evolution affect your classroom teaching?

Dr. Parker: It certainly did. In my early years of teaching at the high school and college levels, I worked hard to convince my students that evolution was true. I even had some creationist students crying in class. I thought I was teaching objective science, not religion, but I was very consciously trying to get students to bend their religious beliefs to evolution. In fact, I had a discussion with high school teachers in a graduate class in which I was assisting to encouraging them to aim at persuading students to adapt their religious beliefs to the concept of evolution!

Interviewer: I thought you weren't supposed to teach religion in the American public school system.

Dr. Parker: Well, maybe you can't teach the Christian religion, but there is no trouble at all teaching the 'evolutionary religion'! I've done it myself, and I've watched the effects that accepting evolution has on a person's thought and life. Of course, I once thought that effect was good – liberating the mind from the shackles of revealed religion and making a person's own opinions supreme.

Interviewer: Since you found evolution so satisfying and enjoyed teaching it to others, what on earth made you change your mind?

Dr. Parker: I've often marvelled that God could change anyone as content as I was, especially with so many religious leaders (including two members of the Bible department where I once taught!) actually supporting evolution over creation. But through a Bible study group my wife and I joined (originally for purely social reasons) God slowly convinced me to lean not on my own opinions or those of other human authorities, but in all my ways to acknowledge Him and to let Him direct my paths. It is a blessed experience that gives me an absolute reference point and a truly mind-stretching eternal perspective.

Interviewer: Did your conversion to Christianity make you an instant creationist?

Dr. Parker: No, at least not at first. Like so many before and since, I simply combined my new-found faith with the 'facts' of science and became a theistic evolutionist and then a progressive creationist. I thought the Bible told me *who* created, and that evolution told me *how*. But then I began to find scientific problems with the evolutionary part, and theological problems with the theistic part. I still have a good many friends who believe in theistic and/or progressive evolution, but for me, after examining it I finally had to give it up.

Interviewer: What theological problems did you find with theistic evolution?

Dr. Parker: Perhaps the key point centred on the phrase, "very good". At the end of each creation period in Genesis chapter one (except the second) God said that His creation was good. At the end of the sixth period He said that all the works of His creation were very good. Now all the theistic evolutionists and progressive creationists I know, including myself at one time, try to fit 'geologic time' and the fossil record into the creation periods. But regardless of how old it is, the fossil record shows the effects of the same things that we have on earth today – famine, disease, disaster, extinction, floods and earthquakes. So, if fossils represent stages in God's creative activity, why should Christians oppose disease and famine, or help preserve endangered species? If the fossils were formed *during the creation week*, then all these things would be very good.

When I first believed in evolution, I had sort of a romantic idea about evolution as an unending progress. But in the closing paragraphs of the *Origin of Species*, Darwin explained that evolution, the "production of higher animals", was caused by "the war of nature, from famine and death". Does "the war of nature, from famine and death" sound like the means God would have used to create a world which He describes as "very good"?

In Genesis 3, Romans 8 and many other passages, we learn that such negative features were not part of the world that God created, but entered only after Adam's sin. By ignoring this point, either intentionally or unintentionally, theistic evolutionists and progressive creationists come into conflict with the whole pattern of Scripture: the great themes of Creation, the Fall, and Redemption – how God made the world perfect and beautiful; how man's sin brought a curse upon the world; and how Christ came to save us from our sins and to restore all things.

Interviewer: If the Bible is that clear, why do so many Christians still believe in theistic evolution and progressive creation?

Dr. Parker: Well, I can't speak for all of them, but I can tell you the problems *I* had to overcome before I could give up theistic evolution. First, I really hate to argue or take sides. When I was a theistic evolutionist I didn't have to argue with anybody. I just chimed in smiling at the end of an argument with something like, "Well, the important thing is to remember that God did it."

Then there is the matter of intellectual pride. Creationists are often looked down upon as ignorant throw-backs to the nineteenth century or worse, and I began to think of all the academic honours I had, and to tell you the truth I didn't want to face that academic ridicule.

Finally, I, like many Christians, was honestly confused about the Biblical issues. I first became a creationist while teaching at a Christian college. Believe it or not, I got into big trouble with the Bible department. As soon as I started teaching creation instead of evolution, the Bible Department people challenged me to a debate. The Bible Department defended evolution, and two other scientists and I defended creation!

That debate highlighted how 'religious' evolution really is, and the willingness of Christian leaders to speak out in favour of evolution makes it all the more difficult for the average Christian to take a strong stand on creation. To tell you the truth, I don't think I would have had the courage, especially as a professor of biology, to give up evolution or even theistic evolution without finding out that the bulk of scientific data actually argues *against* evolution.

Interviewer: In that sense, then, you're saying it was really the scientific data that completed your conversion from evolution, through theistic evolution and progressive creation, to biblical, scientific creationism?

Dr. Parker: Yes, it was. At first I was embarrassed to be both a creationist and a science professor, and I wasn't really sure what to do with the so-called 'mountains of evidence' for evolution. A colleague in biology, Allen Davis, introduced me to Morris and Whitcomb's famous book, *The Genesis Flood*. At first I reacted strongly against the book, using all the evolutionist arguments I knew so well. But at that crucial time, the Lord provided me with a splendid Science Faculty Fellowship award from the N.S.F. (National Science

Foundation), so I resolved to pursue doctoral studies in biology, while also adding a cognate in geology specially to check out some of the creationist arguments firsthand. To my surprise, and eventually to my delight, just about every course I took was full of more and more problems for evolution, and more and more support for the basic points of Biblical creationism outlined in *The Genesis Flood* and Morris' later book, *Scientific Creationism.*

Interviewer: Such as?

Dr. Parker: One of the tensest moments for me came when we started discussing uranium-lead and other radiometric methods used for estimating the age of the earth. I felt sure that all the silly creationistic arguments would be shot down, but just the opposite happened.

In one graduate class, the professor told us we didn't have to memorize the dates of the geologic systems since they were far from certain and riddled with contradictions. Then in geophysics we went over all of the assumptions that go into radiometric dating. Afterwards, the professor said something like this, "If a fundamentalist ever got hold of this stuff, he would make havoc out of the radiometric dating system. So, keep the faith." I was shocked! If it was a matter of keeping faith, I had another faith I preferred to keep.

Interviewer: Are there other examples like that?

Dr. Parker: Lots of them. One concerns the word paraconformity. In *The Genesis Flood*, I had heard that paraconformity was a word used by evolutionary geologists for fossil systems out of order, but with no evidence of erosion or overthrusting. My heart really started pounding when paraconformities and unconformities came up in geology class. What did the professor say? Essentially the same thing as Morris and Whitcomb. He presented paraconformities as a real mystery and something very difficult to explain in evolutionary or uniformitarian terms. We even had a field trip to study paraconformities which only served to emphasize the point.

So again, instead of challenging my creationist ideas, the geology I was learning in graduate school was supporting it. I even discussed a creationist interpretation of para-conformities with the professor, and I finally found myself discussing further evidence of creation with fellow graduate students and others.

Interviewer: What do you mean by 'evidence of creation'?

Dr. Parker: All of us can recognize objects that man has created, whether paintings, sculptures or just a Coke bottle. Because the pattern of relationships in those objects is contrary to relationships that time, chance and natural physical processes can produce, we know an outside creative agent was involved. I began to see the same thing in a study of living things, especially in the area of my major interest, molecular biology.

All living things depend upon a working relationship between inheritable DNA/RNA and proteins, the chief structural and functional molecules. Just as phosphorus, glass and copper will work together in a television set only if properly arranged by human engineers, so DNA and protein will only work in productive harmony if properly ordered by an outside creative agent. I presented the biochemical details of this DNA-protein argument to a group of graduate students and professors, including my professor of molecular biology. At the end of the talk, my professor offered no criticism of the biology or biochemistry I had presented. She just said that she didn't believe it because she didn't believe there was anything out there to create life. But if your faith permits belief in a Creator you can see the evidence of creation in the things that have been made (as Paul implies in Romans 1:18-20).

Interviewer: Has creationism influenced your work as a scientist and as a teacher?

Dr. Parker: Yes, in many positive ways. Science is based on the assumption of an understandable orderliness in the operation of nature, and the Scriptures guarantee both that order and man's ability to understand it, infusing science with enthusiastic hope and richer meaning. Furthermore, creationists are able to recognize *both* spontaneous *and* created (i.e. internally and externally determined) patterns of order, and this opened my eyes to a far greater range of theories and models to deal with the data from such diverse fields as physiology, systematics and ecology.

Creationism has certainly made the classroom a much more exciting place, both for me and my students. So much of biology touches on key ethical issues, such as genetic engineering, the ecological crisis, reproduction and development, and now I have so much more to offer than just my own opinions! Of course, on the most basic matter of origins, my students and I have the freedom to discuss *both* evolution *and* creation, a freedom tragically denied to most young people in our schools today. Creationists have to pay the price of academic ridicule and occasional personal attacks, but these are nothing compared to the riches of knowledge and wisdom that are ours through Christ!

For the Reader

"What worries me is that so many physicists and geologists seem to think that peppered moth or finch beak observations illustrate a mighty creative force that produced moths and birds in the first place."

Professor Phillip E. Johnston
Author of *Darwin on Trial*

In the BBC series 'Great Britons' in 2002, Charles Darwin was listed as the 4th greatest Briton ever to have lived. In his contribution to the series, BBC correspondent Andrew Marr stated: "*We have many local heroes; we have only one world changer. His name is Charles Darwin.*" In a similar vein James Watson, co-discoverer of the structure of DNA, stated, "*Charles Darwin will eventually be seen as a far more influential figure in the history of human thought than either Jesus Christ or Mohammed.*"

Darwin's claim to fame goes back to 1838 when he became convinced of a radical idea: "*species are mutable productions.*" In layman's terms that means there's nothing to stop fish evolving into amphibians. Back in the 19th century, so sacred and universal was belief in the fixity of species that just verbalising his contrary conclusion felt to Darwin like "*confessing a murder.*"

Prior to Darwin, the Bible's statement that living things reproduced "*after their kind*" was unknowingly being taken too far by theologians and scientists. It was widely thought that *every variation* now visible in nature had existed *in its current form* in the Garden of Eden. Against this background, when Darwin observed the variations that dog-breeders were producing, it made him wonder – if man could bring about

41

such changes in so short a time, perhaps over millions of years nature, without God's help, had produced all the various species of plants and animals in the world from a single-celled ancestor.[1]

Then there were the thirteen diversified but similar species of finches Darwin documented on the Galapagos Islands. Rather than God's creating each separate species of finch, Darwin reasoned that all the finch types probably originated naturally from a common ancestor by gradual modification and 'natural selection'. Darwin's theory was sublimely simple: favourable advantageous variations in offspring, naturally selected in the struggle of life over long periods of time, result in the formation of new species. His idea had *some* truth in it, but in positing that organism change was to all intents and purposes *unlimited* and *infinitely* variable, he far exceeded what was scientifically verifiable. (Hardly surprising since Darwin, along with all his contemporaries, was ignorant of genetics).

The fact is, natural selection's effect on finch beak size could never begin to explain where the finches came from in the first place, nor did it prove that all of nature was one great continuum or tree of life. In truth, other considerations, such as the fossil record, stood against Darwin's whole idea. Darwin puzzled, *"Why then is not every geological formation and every stratum full of such intermediate links? Geology assuredly does not reveal any such finely graduated organic chain; and this, perhaps, is the most obvious and gravest objection which can be urged against my theory."*[2]

Still, Darwin decided to set forth his theory, burying his doubts under what he called "the extreme imperfection of the geological record." The *Origin of Species* was a publishing sensation. Almost overnight, the prevailing belief that all the beautiful and ingenious contrivances in nature were creations of God collapsed. 'Design' was merely the 'appearance of design' brought about by 'natural selection'. God, if He existed at all, was clearly redundant – and man, far from being created in the image of God, was a mere cousin of the ape. Small wonder that Ernst Mayr, the 20th Century's most prestigious Darwinist authority called Darwinism *"...perhaps the most fundamental of all intellectual revolutions in the history of mankind."*[3]

So, what is the truth? Did every species in nature evolve by mutation and natural selection from an ancient spontaneously generated 'replicator', or did God create each plant and animal *as we see them today* and place them *in their current location* during creation week? Actually, neither idea is true. It is essential to understand that the 'created kind' mentioned repeatedly in Genesis chapter 1 is often at a higher level of classification than the modern-day species or even the genus. It's quite possible that even some whole 'orders' of animals may have derived from an original 'created kind'. The fact that different species and genera can be interbred – a zebra with a horse, a lion with a tiger and a camel with a llama – proves that, even if their offspring are in some cases sterile, they must have descended from the same original created kind.

The evidence available through, for example, the study of genetics and the fossil record, points to the fact that God created thousands of pairs of basic types in the beginning, each of which possessed the genetic information and flexibility to produce wide variety in its offspring depending on its environment and location. Selective breeding by humans and genetic mutation have also contributed to variation. Chihuahuas, Terriers and Pekinese all have a mutation in the gene for an important growth regulator (IGF1) which results in less production. Whippets with one mutated copy of the myostatin (MTSN) gene and one normal copy end up being more muscled than normal whippets and consequently can run faster. Thus creation scientists do not believe that God created all the species *exactly as we see them today.*[4]

The differences between Chihuahuas and Great Danes are of a different kind altogether to the radical differences between classes like mammals, reptiles, fish, amphibians and birds. (Which is why they were classified as different classes in the first place). Each class possesses a number of unique defining characteristics which are *not found in any other class*. So, the evolution of mammals from reptilian type ancestors requires the development of new features such as mammary glands and a milk supply, a hair covering, a temperature control system, a corti, a diaphragm and a fourth chamber in the heart. Is this kind of change possible by random mutation and natural selection? Sadly for Darwinists the answer is no. Why? There are too many genetic obstacles in the way, even given vast ages of time.

Since mutations are almost always harmful, a major function of genes is to resist mutation, not facilitate it. In a Darwinian world every single mutation must be advantageous or it will be quickly weeded out. The science literature is replete with discussions of other obstacles to mutation-based evolution, from a too slow mutation rate to the fact that mutations are often recessive and prevailed against by dominant genes in reproduction. Then there's the problem of information. Just as turning a single short telegram message into an Encyclopaedia would demand the introduction of thousands of intelligent grammatically correct sentences, so evolving new biological organs and systems demands a huge gain in complex genetic information. By 'information' scientists mean novel, meaningful sequences of DNA. Yet, no mutation known to man has ever led to an *increase* in such genetic information.

Probably the gravest problem with the neo-Darwinian theory - that mutations naturally selected have built all of nature - is the simple fact that such a theory ignores what Michael Behe calls the 'edge of evolution'. Behe is Professor of Biological Sciences at Lehigh University in Pennsylvania. In addition to publishing over 35 articles in refereed biochemical journals, he has written editorial features in Boston Review, The American Spectator, and The New York Times. His book, *Darwin's Black Box*, has sold over 250,000 copies and has been called one of the 100 most influential books of the 20th century. Behe points out in his second book, *The Edge of Evolution* (Free Press, New York, 2007), that newly available genetic data on humans and our microbial parasites (malaria, HIV, *E. coli*) allows us to give a scientific answer to the question, "Based on hard empirical data, what can Darwinian evolutionary processes actually do?" The answer? Astonishingly, even under intense selective pressure, and given an astronomical number of opportunities, random mutation and natural selection yield only *trivial*, mostly *degenerative*, changes. For example, selection of a random mutation in the malarial parasite explains its development of resistance to chloroquine, but only because it has a huge population and a short life-cycle. That won't work for large, complex creatures with smaller populations and longer generations.

Not that it hasn't been tried. Eminent evolutionary geneticist Richard Goldschmidt bred gypsy moths for 20 years and a million generations. All he ever produced were gypsy moths. Famed American plant breeder Luther Burbank admitted that though he could breed a plum anywhere from ½" to 2½" long (1.3cm to 6.3cm), he could neither go as small as a pea nor as big as a grapefruit. As biology Professor Lane Lester (PhD in genetics) and Ray Bohlin (PhD in molecular biology) have observed, "*Natural selection, recombination, mutation and speciation can all interact in concert to bring about startling variation within the created prototype* [kind]...*but there are limits to biological change.*"5

All of this goes to show that the Bible's ancient claim that plants and animals reproduce only 'after their kind' stands as a scientifically sound statement, provided we understand that being in the same 'created kind' means descending from the same ancestral gene pool. However, Darwin's claim that all of life, including humans, came from a single-celled organism in a little warm pond millions of years ago by nothing more than accidental mutations selected by nature, collapses through lack of evidence.

Granted, the vast majority of scientists still believe in Darwinism partly due to lack of exposure to the other side's arguments, and partly due to an *a priori* commitment to naturalism. Richard Dawkins is typical when he states: *"I'm a Darwinist because I believe the only alternatives are Lamarckism or God, neither of which does the job as an explanatory principle."*[6] Or take Professor D.M.S. Watson writing in Nature magazine; *"Evolution [is] a theory universally accepted not because it can be proven by logically coherent evidence to be true, but because the only alternative, special creation, is clearly incredible."*[7] Distinguished Harvard geneticist, Richard Lewontin, is even more forthright. Though speaking of astronomy at the time, his comments are revealing: *"We take the side of science in spite of the patent absurdity of some of its constructs, in spite of its failure to fulfil many of its extravagant promises of health and life, in spite of the tolerance of the scientific community for unsubstantiated just-so stories, because we have a prior commitment, a commitment to materialism. It is not that the methods and institutions of science somehow compel us to accept a material explanation of the phenomenal world, but, on the contrary, that we are forced by our a priori adherence to material causes to create an apparatus of investigation and a set of concepts that produce material explanations, no matter how counter-intuitive, no matter how mystifying to the uninitiated. Moreover, that materialism is absolute, for we cannot allow a Divine Foot in the door."*[8]

The Bible calls Lewontin's type 'willingly ignorant'. Yet with atheistic Darwinism further away than ever from providing an explanation for the origin of the universe, life and consciousness, honest open-minded enquirers have at least three pathways open to them each of which leads to belief in the existence of God.

Consider the following:

1. The Information Pathway

Just as a single intelligent radio signal from outer space would indicate the presence of intelligent life 'out there', so the digitally coded DNA database in the cell evidences a designer and creator. Complex specified information cannot self-assemble – it demands a designer.

2. The Elimination Pathway

There are only four possibilities for the cosmos' origin:

a. It came from nothing accidentally
b. It came from nothing supernaturally
c. It has always been here
d. It is not really here; it is an illusion

The only option that will stand up to scientific scrutiny is option b. Option a is out, since from nothing, nothing comes. Option c is also out due to the second law of thermodynamics. Would your watch still be ticking if it had always been here? Option d is plain silly. One has to exist even to be able to deny one's existence. Option b must be the truth. Since whatever begins to exist has a cause, and the universe began to exist, the universe must have had a cause. Yet, prior to the beginning of the universe there was neither matter, energy *nor time*. Thus the cause of the universe is timeless and eternal. This renders the question "Who made God?" redundant. He is the uncreated creator, the uncaused causer, the unmoved mover without whom nothing else could ever have existed.

3. The Revelation Pathway

There are numerous ways in which God has revealed Himself to mankind. From the mighty galaxies to the nano-machinery of the cell, His wonderful works reveal a powerful, intelligent creator and designer. Added to creation's revelation is the written revelation of the Bible. The predictive accuracy, scientific accuracy, medical accuracy, uniqueness, durability, harmony and life-changing power of the Bible give evidence that it is indeed the very word of God. To cap it all God has given us His incarnate revelation – He has spoken to us through His own Son, the Lord Jesus Christ.

God, the author of the laws of science, is also the source of moral and spiritual absolutes, a fact reflected in the conscience of human beings who all know, for example, that it is always wrong, at all times in all places to torture little children.[9] That is, absolute right and wrong is independent of society and history. Since every law has a law-giver and there is a moral law, there must be a moral law-giver. Since the beginning of history the human race has rebelled against its moral law-giver and Creator. It is an integral part of fallen human nature to play at being God and look for excuses neither to acknowledge nor to be accountable to Him. Scientific materialism, based on Darwinism, is but one of those excuses.

The fact is, God is holy - and despite our best efforts we have all fallen short of His perfect standard of righteousness. Though we like to think we are basically good, when we look into the mirror of God's law we find corruption within. Have you always put God first in your life, or have other ambitions and possessions come first? Have you ever told a lie? Have you ever stolen anything? Have you ever lusted after another's spouse? Scrutinized by the holy laws of God on judgment day, would you be found guilty or innocent? According to the Bible, the whole world is under condemnation for breaking God's laws (Romans 3:19). Due to the fact that we not only have committed acts of sin, but are sinners by birth, being connected to a fallen human race, we cannot put our relationship right with God by our own self-efforts or good works. The first step on the way to becoming right with God is to come to a proper sense of the dreadfulness of our sin against God, and to cast away any thought of self-righteousness and self-sufficiency. We have to admit we are lost and need a Saviour.

God saw our impossible situation. Though He can never lower the standards of His righteousness and overlook sin, out of divine love and compassion He became man in the person of Jesus Christ. He lived a perfect life among us and willingly went to death on a cross where he offered Himself as a sacrifice for the sins of the world. In God's sight our sin is so serious that it must be punished, either in hell where the unrepentant sinner will receive the eternal judgment he deserves, or in the substitutionary death of Christ on the cross. Three days after dying Jesus Christ broke the power of death and rose physically from the dead. After this He ascended back to heaven where He sits as God and Saviour. Today the call comes to you personally – turn from your sin, let go of any former confidence and trust alone in the Son of God, Jesus Christ, resting on His finished sacrifice for your deliverance from sin and hell.

If through this book you have trusted in the Lord Jesus Christ alone for your eternal salvation you should immediately take the following steps:

1. Thank Him for what He has done for you and ask yourself the question, "What can I now do for Him?"
2. Start speaking daily to Him in prayer from your heart, bringing Him praise and thanksgiving, as well as asking Him for blessings.
3. Obtain a Bible and start reading and studying it. It's best to begin with the Gospels (e.g. Mark or John) and read through the New Testament first before progressing to the Old Testament. Ask God to give you understanding as to how to apply the Bible's teachings practically to your life.
4. Find a Bible-believing Church and regularly attend its meetings and be baptised.
5. Tell others what the Lord Jesus has done for you.

Notes:

1. The effects of selective dog breeding remain an important icon of neo-Darwinism. While it is true that turning some existing genes or regulatory elements on or off, or changing them slightly by simple, single mutations, can certainly affect the shapes and other properties of dogs to a degree, that does not explain where the complex systems controlling the dogs' development came from in the first place. Furthermore, these types of genetic changes are of a wholly different kind than is needed for bacterium-to-Beethoven style evolution.
2. Charles Darwin, The Origin of Species, (London, Penguin Classics, 1985), p. 292
3. Ernst Mayr, Science, June 2, 1972 p. 981
4. http://www.creationontheweb.com/content/view/271
5. Lane Lester & Raymond Bohlin, The Natural Limits To Biological Change (Dallas, Probe Books, 1989), p. 175-6, 14
6. Richard Dawkins, "A Survival Machine." In The Third Culture, edited by John Brockman (New York, Simon & Schuster, 1996), p. 75-95
7. Prof D. M. S. Watson, Adaptation, *Nature*, 1929, 124:233
8. Richard Lewontin, "Billions and billions of demons", (The New York Review, January 9, 1997), p. 31
9. The Bible, Romans 2:15: "…they show the work of the law written in their hearts, their conscience bearing witness, and their thoughts alternatively accusing or else defending them…"

Acknowledgments:

1. **The Challenge of Creation**, taken from *On the Seventh Day*, edited by John F. Ashton PhD, Master Books © 2002, p. 33-38, and from *In Six Days*, p. 239-245, edited by John F. Ashton, New Holland Publishers (Australia) © 1999 by kind permission.

2. **Testing Truth with an Open Mind**, taken from "Approaching the Bible with an Open Mind," in *Scientists who find God*, edited by Eric C. Barrett PhD, foreword by Dr. Bob Provost, p. 10-20, Slavic Gospel Association © 1997 by kind permission.

3. **From Oxford Atheist to Leading Creationist**, taken from his autobiography, *Fulfilled Journey*, with his wife Beate Wilder-Smith, Word for Today Publishers, USA © 1998 by kind permission.

4. **Superb Design**, from *On the Seventh Day*, p. 87-93, edited by John F. Ashton PhD, Master Books © 2002 by kind permission.

5. **From Evolution to Creation**, taken from *Evolution to Creation*, impact No. 49, July 1977, taken from the Institute for Creation Research website © 2004 by kind permission.

Also available:

Dawn of the New Age	*5 New Agers Relate Their Search for the Truth*
Angels of Light	*5 Spiritualists Test the Spirits*
The Pilgrimage	*5 Muslims Make the Greatest Discovery*
Witches and Wizards	*5 Witches Find Eternal Wisdom*
They Thought They Were Saved	*5 Christians Recall a Startling Discovery*
Messiah	*5 Jewish People Make the Greatest Discovery*
Light Seekers	*5 Hindus Search for God*

If you would like confidential help or further information, please feel free to contact us. We can supply literature and details of Bible believing churches in your area. If this book has been a help to you please let us know.

Published by:
Penfold Books
P.O. Box 26, Bicester, Oxon, OX26 4GL.
Tel: +44 (0) 1869 249574
Fax: +44 (0) 1869 244033
Email: penfoldbooks@afo.net
Articles & News: www.webtruth.org
Evolution: www.atheistdelusion.net
Books: www.penfoldbooks.com

Copyright: Penfold Books 2008
ISBN: 1-900742-24-1